The DNA of Selling

What You Won't Learn in Business School

By Gerry Shaltz

iUniverse, Inc.
New York   Bloomington

iUniverse books may be ordered through booksellers or by contacting:

iUniverse
1663 Liberty Drive
Bloomington, IN 47403
www.iuniverse.com
1-800-Authors (1-800-288-4677)

Because of the dynamic nature of the Internet, any Web addresses or links contained in this book may have changed since publication and may no longer be valid. The views expressed in this work are solely those of the author and do not necessarily reflect the views of the publisher, and the publisher hereby disclaims any responsibility for them.

ISBN: 978-1-4401-7957-0 (sc)
ISBN: 978-1-4401-7959-4 (dj)
ISBN: 978-1-4401-7958-7 (ebook)

Printed in the United States of America

iUniverse rev. date: 03/19/10

Dedication

This book is dedicated to my wife, **Sharon.** Her love and boundless faith in me are the key ingredients that made this undertaking possible.

Gerry Shaltz
Author, Lecturer, Entrepreneur

Comments from Graduating MBA students of 2007, Anderson School of Management, UCLA

"...one of the best and most practical seminars I've attended ... ever."

"Sales is not something that can be completely taught... I feel it's half inherent ability and half experience. Gerry translates nearly all of his years of experience into tips and ideas that are applicable to all levels of one's career. ... He gets you halfway up the hill before you make a single call."

"Gerry is a dynamic speaker! He converts his extensive sales experience into a series of easy-to-remember, easy-to-implement helpful hints that will prove useful for anyone... whether they're in the sales business, or just needing to effectively influence people."

"A no-nonsense, easy-to-follow guide for both beginners and seasoned professionals in the business of selling."

"An exquisite blend of inspiration, humor, and straight talk."

" Practical, encouraging methods.... Contains crucial information that serious sales professionals need to succeed. Provides all the right tools of the business."

"Gerry's tips and ideas are coupled with vivid stories from his years of experience. ... this makes them easy to remember and to pull from when the situation calls for it!"

Here's what others are saying about author, teacher, mentor, entrepreneur, Gerry Shaltz:

Dear Gerry,

Mark Twain was asked if he could write a three-page story in three days. He replied, "I'll give you a 30-page story in three days, but it will take me 30 days to write a three-page story."

I can only imagine how difficult it was to cram a lifetime of learning into two hours. Your performance was nothing short of remarkable. The applause you received was better than most professors get after teaching an entire quarter of classes. People are still talking about it. In fact, I retold the "Librarian" joke earlier today to give people a taste of what they missed and to magnify their fear of loss at not participating in future sales modules. THANK YOU FOR EVERYTHING!

Drake Watten
President, UCLA MBA graduating class of 2007

Contents

*"Gerry is the best pure salesperson I have known in my career. His
natural talent combined with hard work and a desire for continuous
learning allowed him to take his career to a level of excellence few will
achieve. This is someone you can definitely learn from."*

Mike West
Former President/COO, Octel Communications
Former Chairman, Vina Technologies
Former Chairman, Extreme Networks

Acknowledgements

I would like to acknowledge Sandy Martin, who guided me through the daunting task of writing my first book. I didn't make it very easy for her. Her extensive literary experience, infinite patience and gentle spirit helped me every step of the way.

My associate and long time friend, Dale Penn, (writing his first book as I wrote mine), eagerly shared resources with me, propped me up when I was down and was always there when I needed advice and support. Dale is among the sales greats I refer to in this book.

I want to thank Bob Zider, president, the Beta Group, past president of the Harvard MBA class of 1976, for encouraging me to write this book. He, along with Drake Watten, Boston Consulting Group, past president of the UCLA MBA graduating class of 2007, were directly responsible for making it possible for me to lecture on selling for the UCLA, Anderson School of Management, graduating MBA class of 2007. This was the final event that motivated me to launch the writing of this book.

I would also like to acknowledge Dave Plough, Stanford MBA, president & CEO, Portaero, Inc., for appreciating my selling skills. It was his endorsement of me that resulted in a sales seminar for a group at Stanford University – a valuable experience among some very bright individuals.

I can't forget to acknowledge Larry Goldfield, who stuck his neck out many years ago to help me land my first real paying sales job. He was instrumental in lifting me up several income levels to a new and exciting career in professional selling.

A big "Thank you!" to Sandy Bondan of the Beta Group, who has never said "No" to the many requests I've made for help over the years. She's truly a talented team player with a wonderful attitude. She labored diligently for many hours transcribing the video of my UCLA lecture so that I would have a base guide for writing the book.

I want to express my gratitude to George Dickson, III, who, over the past fifteen years redefined, by example, the words endurance, perseverance, tough-mindedness and faith. George never ceases to amaze and inspire me. He is a valuable business associate, a loving and loyal friend, who believes that I have been his mentor. In reality, he has been mine. George is one of the sales giants I've studied. His image frames many of the lessons detailed in this book.

I also want say "Thank you" to a small group of talented individuals who helped me with this book in many different ways. Beth Strahle did a wonderful job editing and proofreading, a tedious challenge to say the least. Jim McAllister, World Vision International, was masterful in coordinating the production of this book. Rebekah Roose pieced together a puzzle of words and pages, formatting and designing the graphics. Greg Flessing, Fresh Air Media, was so generous with his time and talent helping with the title and sharing numerous, creative ideas to enhance the visual impact of the book. Harold Ross, with his eloquent use of the English language, bailed me out of several serious word holes that had stopped me cold. He also came up with several ideas to improve the book cover. A "Thank you" to Mike West for encouraging me to tell several personal selling stories, insisting that doing so would add flavor to the book. David Glenwinkel, Outcomes for Business, the master of keeping record numbers of plates spinning in the air at the same time, was accessible and supportive. His insistence that this book was worth completing added the extra fuel I needed to do so.

Thank you to my daughter, Kayla Beth Shaltz, whose presence in my life has been one of the motivating forces that kept me working hard and trying to be a good role model.

Inder Jaisingh got me back on track with his humor and encouragement each time I wanted to chuck the entire project, "Hey man, don't be a wuss, you're almost done." He was right.

About the Book

The DNA of Selling brings a wealth of knowledge and experience the author gained during many years of building a lucrative and fulfilling career in sales. Gerry Shaltz has compiled his most powerful sales tactics into this easy-to-follow guide, complete with step-by-step instructions. He graciously shares his methods in ways designed to meet business and sales professionals at every skill level.

Readers will find crucial tools needed to gain, renew, or enhance their own selling skills. This book will also build confidence in students aspiring to create successful careers in selling and business at every level. Even sales veterans working to stay at the top will find inspiration here.

The author's seminars and lectures have been widely requested since he first began revealing his trade secrets. He shares his wisdom and proven success-building strategies with numerous businesses, organizations and university business schools such as UCLA, Anderson School of Management, and others.

These strategies were developed over literally, thousands of sales presentations. In this book, Gerry Shaltz clearly and openly presents tools that many experienced business and sales experts wish they'd had when they first started out.

"Most business school full professors have never been businessmen. They got their master's, PhD, then taught and wrote their way up the ladder to tenure. There is no tenure in business. If you can't sell (your product, your service, yourself) you are gone. Sadly, most business professors feel that teaching selling is "beneath" them. You might succeed in business school without selling but you can't succeed in the business world without it. Imagine a medical school where the teaching staff has never performed surgery themselves, yet they are instructing the interns and residents. That's what you have at business schools."

Bob Zider, *President, The Beta Group*

"I find it useful to remember that everyone lives by selling something."

Robert Louis Stevenson

Introduction

How it All Began

I knew early on that I'd never be an NBA star or score a touchdown for the 49ers. I also knew that I was good at persuading people to do what I wanted them to do. I found my calling in the business of selling as a ten-year-old boy knocking on doors in central Los Angeles. It was then that I closed my first sale while soliciting newspaper subscriptions.

The newspaper company paid me seventy-five cents for each subscription I wrote. They paid me for simply convincing people to have a newspaper delivered to their doorstep rather than buying it at the newsstand. I was selling, and I couldn't believe how easy it was to earn money that way. This realization abruptly ended my lawn-mowing career. It was a great beginning, and I've been working in sales and associated endeavors ever since.

In time, I came to learn that no job in itself offers "job security." Ultimate security is based upon your ability to produce. This is one of the many reasons I was drawn to straight commission selling. It provides an entrepreneurial business environment whereby one is really in business for oneself, so that "you earn what you get and get what you earn." I also appreciated being somewhat isolated from the distractions, disappointments, and tedium of corporate politics. Top salespeople are always in demand.

What You Won't Learn in Business School

Each year, business schools across America send thousands of MBA graduates out into the business community without having offered a single course in selling. Selling doesn't exist in the curriculum. These grads set out on an expedition into an unknown terrain. The one true certainty is that they will need selling skills in order to increase their chances for success.

They are enthusiastic, filled with information and confidence. Most believe that they are ready to go out and "slay the dragon." They feel certain that their education has given them what they need to be successful in the business arena.

Business schools offer classes in marketing. Many believe that marketing and selling are synonymous and that selling is merely a sub-component of marketing. The two, by definition, are not the same. When my associate, Dale Penn's title was changed from VP of Marketing to VP of Sales, he found it necessary to come up with simple definitions to define the core differences between the two: **"Marketing uses money to generate ideas. Sales uses ideas to generate money."**

Once satisfied with his definition, he was able to go forward in his new position with sufficient clarity. He became highly successful as an outstanding leader of salespeople. It took talent, selling knowledge, the practice of sound selling tactics, and hard work to achieve the pinnacle sales position in his industry.

According to the Random House 2006 *Unabridged Dictionary,* the definition of marketing is: "The total activities involved in the transfer of goods from the producer or seller to the consumer or buyer, including advertising, shipping, storing, and selling." I believe that the major goal of marketing is to create product awareness. From the big picture view, marketing is strategic – selling is tactical.

Effective selling takes skill and courage. It is fundamental to business at all levels, yet, *business schools don't teach their students selling skills.* They leave it up to the graduating students they "educate" to go out and learn it on their own. Therefore, most of the graduates know little or nothing about the business of selling. Without professional selling skills, most are unprepared to meet the important day-to-day challenges they will face.

During the Q&A section of a lecture I gave to graduating MBAs, I learned that most of the students didn't know how to effectively set an appointment with a decision maker or even how to get past the gatekeeper. Most knew little or nothing about handling objections or how to identify a hidden objection. Few could define a closing question. I asked one group of MBAs to give me an example of a trial close and received no response from an audience of seventy.

Many succeed despite this lack of instruction; others fail because of it. Recently, I spoke with one of the brightest MBA's in his graduating class – a born leader. He said, "I don't even know how to prepare or give a sales presentation. How can I manage salespeople when I don't know anything about selling?" *Business Schools Don't Teach Selling.*

The Stigma of Selling

Many people, especially in the United States, don't trust salespeople. They believe that selling is a lowly profession. This is understandable considering those who have had negative experiences dealing with overly aggressive salespeople.

Some of the stereotypical, negative images are the high-pressure used-car salesperson, the unethical real estate agent, the deceptive insurance representative, or the door-to-door peddler. This is why most companies give their sales personnel titles like Marketing Manager, Product Consultant, Territory Manager and so on. Most consumers would prefer to meet with a Product Consultant than with a Sales Representative.

While it is true that this book mostly benefits people involved in sales, the contents can benefit all business professionals. Selling, by definition here, is to influence or persuade others to do what we want them to do, or to convince them to see our point of view. The importance of selling cannot be overemphasized. Selling is an integral thread woven throughout the entire fabric of every business. Effective selling skills are essential for everyone in every type of business. Why? Because everyone sells, no matter how smart, talented, or educated. No one is exempt from the act of convincing others of something. We all sell.

Top salespeople get noticed. Performance is easily measured against expectations or quotas. One either achieves the objective or they don't. Most people in business are measured by some standard that has a number attached, whether it's a timeline, deadline, production line, or profit line. From the CEO to the salesperson and everyone in between, we are all assigned expectations and are expected to meet them. Production touches every part of a company, and those who sell are on the front line. Remove sales and there is no production – and therefore, no company. Top salespeople are critical to the survival and success of a business – this is why they get noticed.

> *"Your ability to negotiate, communicate, influence, and persuade others to do things is absolutely indispensable to everything you accomplish in life."*
>
> Brian Tracy

Shadow the Aces (Learn From the Best)

I know a great deal about why people succeed or fail in selling. I made it my business to study top salespeople, and had the good fortune of working alongside some of the best. These people, some earning high six-figure incomes or more annually on straight commission, became my teachers.

After identifying top salespeople, I would shadow them as much as possible. Often, I'd ride along to watch and listen to them while they set appointments, gave presentations, and closed sales. It felt good to be in the trenches with the real pros, the highest achievers. I was a sponge soaking up everything, whether or not I understood it. I sometimes emulated their speech patterns, gestures, pauses, and even a few strange quirks. I had little idea of just how valuable it was to mirror these "heavy hitters," but I did it and it worked. It wasn't long before I was producing at a respectable rate. Eventually, I came to understand the logic behind what the top producers were doing. I refined my selling techniques to suit my own personality and preferences. I kept the methods I could own and that worked for me and discarded ones that didn't. Even today I never pass up an opportunity to observe and question top salespeople. It's a continuing curiosity that I acquired early in my career. I'm still learning from the top producers.

What You Will Gain

This book is rich in success-building strategies developed over thousands of sales presentations. It reveals (in detail) the habits, traits, and attributes that truly great salespeople have in common. You'll learn why some salespeople fail and how to avoid their mistakes. You will find information here that is sure to speed up your learning process. These pages reveal secrets that will help you 1) effectively set appointments with decision-makers 2) shorten the selling cycle, and 3) hold the prospect's attention throughout the entire selling process.

Among other things, you will receive a wealth of simple, step-by-step, instructions, proven selling principles, and time-tested tactics that you can confidently place in your toolbox. This toolbox will help you avoid costly mistakes, maintain control of the process, and greatly increase your chances for a successful outcome.

Bill Jensen, author of *Simplicity,* writes, "Focus on the tools and resources that are to be used. Why? Because that's how people get the work done. Because even with a shared vision, more employees than you may care to admit don't have the tools or training to do what's expected of them. Clarity about the use of tools and adequate levels of training is critical, but clarity comes second. Investing in the tools comes first."

My vision is that this book will bring about a greater awareness of and respect for the profession of selling. By focusing on the importance of selling, I believe that the barriers existing today in business schools can and will be removed.

While many educators feel that there isn't sufficient scientific data existing to support selling as a course of study, I believe that the individuals who influence and create curriculum for those institutions need to take this subject seriously. There must be a realization that salesmanship is not beneath the image and dignity of traditional academia. Remember, salesmanship takes place even between a prospective student and a graduate school. Selling is part and parcel of the entire human experience of communication and interaction.

There is a plethora of excellent literature available on sales-related subjects. Among these are studies about "visualization," "Neuro Linquistic Programming (NLP)," "Cognitive Economics," numerous, highly informative books about selling, and courses offered by the Dale Carnegie organization and many others. Those who earn professional incomes selling on straight commission understand that selling is a science unto itself and deserves to be studied. It is present in all aspects of business and our lives.

At some point, I am confident that business colleges will begin to provide sales training classes as a fundamental part of preparing students to become more productive in the world of business. In doing so, business schools will improve the quality of the education that they provide, an educational experience that is more in tune with the realities of day-to-day business. These students will be better prepared to serve their companies, their clients, and themselves.

"I read, I forget. I see, I remember.
I do, I understand."

Chinese proverb

Some useful advice for the reader

It's natural to become enthusiastic about the new and innovative things we hear during a lecture or read in a book. We may get excited and believe that the knowledge itself is going to take us to a higher level of performance. Sadly, many of us realize weeks later that we haven't added or changed a thing. At best, the experience had a short-term effect that soon faded. In reality, it was little more than entertainment.

My advice to you is to begin using the lessons presented here immediately. Start by selecting one key point and commit to begin using it now. I urge you to highlight, underline, write in the margins. Do whatever it takes to help you make the commitment to actually **DO** something. Remember, "If nothing changes, nothing changes."

In summary, information is not enough. It's the application that makes things happen. It is application that completes the circle of learning. One of my heroes, Warren Buffet, said, "If past history was all there was to the game, the richest people would be librarians."

"Motivation is a fire from within. If someone else tries to light that fire under you, chances are it will burn very briefly."
Dr. Stephen Covey

LESSON 1

Habits and Attributes of Sales Greats

"No one lives long enough to learn everything they need to learn by starting from scratch. To be successful, we absolutely, positively have to find people who have already paid the price to learn the things that we need to learn to achieve our goals."

Brian Tracy

Shadow the Aces

In the Introduction of this book, I mentioned that I have worked with some sales giants – professionals who

---

earn mid to high six figures on straight commission. I've had the privilege of working with many of these top salespeople in a variety of industries.

This section is devoted to identifying the commonalities of the most successful salespeople I've known. Obviously, they all excel at *organization, hard work* and *follow-up* – these are the fundamentals for success in any field. However, these gifted individuals are blessed with additional habits and attributes.

Visit any company with a sales staff, and you'll most likely find a couple of "heavy hitters" and the "others." The eighty/twenty rule pretty much holds true. Twenty percent of the people earn eighty percent of the income. You might ask, "What separates the top earners from the others?"

The habits and attributes below will demonstrate what separates the 20 percent top earners from the others.

(1)   They are organized.

(2)   They work hard.

(3)   They follow up.

(4)   They are enthusiastic.

(5)   They are excellent listeners.

(6)   They make a positive first impression – build rapport.

(7)   They are persistent.

(8)   They set goals.

(9)   They engender trust.

(10) They are driven to be the best.

(11) They have a "do it anyway" mentality.

(12) They actively work to improve their skills.

(13) They sell value – not price.

(14) They anticipate objections.

(15) They utilize "tie downs".

(16) They say exactly what they mean (great communicators).

(17) They differentiate themselves from the competition.

(18) They ask for the order.

** Consider the first three points (organization, hard work, follow-up) as a three legged stool. If one leg is removed, the stool cannot stand on its own.

(1) They Are Organized

> *"If you want to make good use of your time, you've got to know what's most important and then give it all you've got."*
> *Lee Iacocca, philanthropist, former Chrysler CEO, former Ford president.*

The sales greats invariably plan their work well in advance. Then they faithfully and consistently work their plan. They have already created an organized list of their priorities because they understand what's important and the value of their time, especially during productive selling hours.

They understand the relationship between Return-On-Investment (ROI) and each activity associated with the selling process. During regular selling hours, you won't find them telling war stories with fellow associates or otherwise wasting time. They understand that it's not a popularity contest among their associates. They get paid for selling. They are prospecting, setting appointments, giving presentations or following up – the core activities that make sales.

(2) They Work Hard

"The harder you work, the harder it is to surrender."

Vince Lombardi

Working hard is a must, but it isn't enough. You must also "work smart." For example, someone might work very hard for weeks setting up a filing system during productive selling hours, but this would cause them to lose money. You don't get paid for your filing system. You get paid for making sales. Filing can be set up at home, in the evenings and on weekends.

During productive selling hours, your energies must be consistently focused on the activities that make sales. The greatest salespeople I know always expect a positive outcome. They visualize it, anticipate it, believe it, and expect it.

There is a constant awareness that asks the question: "Is what I'm doing at this moment the best use of my time?" You won't find great performers wasting time. They're working. They are relentlessly focused on keeping their pipeline full with new prospects, appointments, and presentations. They know that if they don't, nothing will be coming out of the other end of the pipe. This is what successful selling is about.

(3) They Follow Up

"Great salespeople are not born or made. They evolve over time based on their dedication to excellence, and their willingness to serve."

Jeffrey Gitomer

Follow-up with prospects and clients is as important as getting the initial appointment.

You should know that thirty percent of prospects will buy from you if you follow up: I can't count the number of times I've closed sales where the prospect told me that my competitor had given a good presentation but never followed up, not even a phone call. I believe that many of those sales came to me by default. The competition essentially left the field, forfeiting the sale. I was in the right place at the right time, proving myself worthy of the business. The competitor did much of the selling for me. I simply picked up the ball that had been dropped and ran for a touchdown.

"Promise big. Deliver big."

H. Jackson Brown, Jr.
Life's Little Instruction Book

What is "follow-up"? It is doing everything you say you're going to do and then some. It's delivering whatever you're supposed to deliver. It's sending "thank you" letters[1] emphasizing the benefits of your product or service. It's showing up more than once, if necessary, to earn their business. It's delivering better and earlier than agreed. It's getting in front of the decision maker as many times as you can until he or she says, "Yes." It is also consistently thinking and acting in the best interest of the prospect and the sale until it's closed. **Follow-up until the sale is finalized.**

Then follow-up some more to make certain that the clients receive everything promised, on time and in excellent condition. Make certain that training is provided if promised or necessary. Ask the clients if they're happy. If they're not happy, do whatever is reasonable and necessary to make them happy.

The only thing you can do for the customer after the sale is to provide excellent service. This will insure longevity in your business and in your career. A sale is not a sale until the customer is satisfied. The day of "hit and run" sales is long gone.

---

1   Short, handwritten notes are far more personal and more likely to be read and remembered.

> *"Always do what you say you are going to do. It is the glue and fiber that binds successful relationships."*
>
> Jeffrey A. Timmons

(4) They Are Enthusiastic

Enthusiasm is contagious

There is no substitute for enthusiasm, not even intelligence. I managed a salesman who we almost didn't hire because he couldn't pass a simple intelligence test. "Mary had six apples and gave Johnny two. She ate one herself. How many apples did she have left?" He couldn't pass the test. But, he was bubbling with genuine, sincere enthusiasm. He was eminently successful. He also worked hard, was organized and followed up.

Be enthusiastic about your product and the company with which you're associated. Recently, I took bids to have some major repairs done to my driveway. Because our driveway is quite long, I knew the work would be expensive. I met with salespeople for each of the three, competing paving companies. One of them made it a point to share his enthusiasm for the company that employed him. I clearly remember his words. "I have a great job. I really love working for these people. They do things right and they treat us with respect and appreciation." He didn't present the lowest bid, but he got the job. The company he worked for got my business because of him.

People can sense the positive, enthusiastic fervor of a salesperson. It may be the key reason that the prospect says "Yes" to your presentation. Don't assume you can hide a lack of enthusiasm. It is visible in many ways. People can detect your level of conviction based on several things, including the energy and body language you exhibit.

Caution: Fake enthusiasm is worse than none.

(5) They Are Excellent Listeners

"The wise man hears one word and understands two."

Yiddish Proverb

Having the patience to listen with interest and truly hear what a person is saying and feeling is a rare talent, a great virtue. Most of us are thinking about the next thing we're going to say and miss much of what a prospect is saying. Too often, we simply can't wait to take the floor and have our say. Sometimes we interrupt before they've finished speaking. Sometimes we are rude and we lose the sale because of it.

Facilitative listening is about being actively involved with what the other person is saying. At best, it's hearing what the person is not saying; what may be hard for him/her to say. This skill can be learned. It takes keen focus and great patience. I do know that the sales greats have it. If you give prospects enough time, enough sincere interest, enough space and encouragement, they will tell you how they want to be sold. **LISTEN!**

(6) They Make a Positive First Impression – Build Rapport

"There's a difference between showing interest and taking interest."

Unknown

First impressions are lasting and critically important. The prospect's mind automatically shifts into an "assessment mode" the first moment you come into view. This is an involuntary phenomenon that is controlled by the survival instinct.

The mind knows that in order for it to stay alive it must keep the body alive. It looks for inconsistencies. It asks, "What's on the other side of that wall and under that rock?" "Better be careful until I can check this person out." The mind concludes with a semi-permanent message once it has completed its initial "inspection." This process usually takes a few seconds to complete. The conclusion becomes somewhat indelible and very difficult to reverse, particularly if it's negative.

Your mind, as the seller, is also checking the prospect out at the same time. You couldn't stop this process –

even if you tried. What can you do to greatly increase the chances of making a positive first impression?

"A smile is a passport that will take you any place you want to go."

Unknown

(a) Upon meeting, smile warmly, with good eye contact.

(b) Shake hands firmly while continuing to smile and looking the prospect in the eyes. A weak, limp handshake is an immediate turn-off. Squeezing too hard as if it's a contest of strength sends the wrong signal, as well.

(c) Look for opportunities to build rapport. Maintain positive body language and good posture. Sit erect, leaning slightly forward in your chair to demonstrate interest.

(d) Use the person's name frequently.

Build Rapport

"The sales manager of a large New York concern once assembled his sales force to meet J.B. Iden, a leading Broadway director. He had engaged Iden to teach his sales folk how to smile. Iden took 500 people one by one and rehearsed their best smiles and pointed out errors, criticized them and embarrassed them. Many salespeople thought they knew how to smile because they smiled every day on the sales firing line, but this specialist in communicating human emotions said what they thought were smiles were smirks. A smile wins good will, a smirk destroys it, and the eyes make the difference. In a true smile the eyes also smile. The eyes are warm. In a smirk only the mouth does. You need to make note of this and think about it. The eyes may continue to be hard, harsh, and unfriendly. Can you imagine spending two weeks learning how to smile? This really happened. These salespeople went out and increased sales by 15%. The right kind of smile will make your job surer and more effective and will help you close sales."

From the book
Secrets of Closing Sales *by Roth and Alexander.*

One of the worst things a salesperson can do is to push a business card at the prospect prior to starting his presentation. I liken this to going out on a first date with someone you're very attracted to. Immediately, upon meeting, you throw a bucket of ice water in their face, and then ask, "Your place or mine?"

People like to do business with people they like and trust. It is crucial to build rapport prior to presenting your product. Building rapport is about finding a common anchor with the prospect.

Lord Nelson said, "I owe all my success in life to having been always a quarter of hour before my time." When you show up fifteen minutes early, you might hear or see things you wouldn't be privy to otherwise. You might have a friendly conversation with the receptionist or possibly another employee. Look for items of interest in the waiting area that could be utilized to initiate a conversation with the prospect. Perhaps the prospect recently won a golf tournament, or perhaps his child did something notable. When you walk into that person's office, look for collectables, hobbies. Look for family photos or trophies.

Whenever I notice happy, busy employees, one of my favorite things to do is to walk in with enthusiasm and say, "I visit many businesses every week, and it's easy to notice when people are happy, energetic and busy. This is an excellent indicator of a successful business with good leadership. There's a good vibe here. How did you create this kind of environment? How do you find people like that?" The point here is to find something of interest that the prospect can feel good talking about. Give the prospect the opportunity to tell you about the "big fish" story or how the business was started.

Smile, maintain eye contact, and be patient. Use the prospect's name frequently. People love to hear their name. The better your rapport is with the prospect, the better your chances are of earning his business. Don't be too eager to commence with your presentation. Build rapport first if possible.

(7) They Are Persistent

Sincere, respectful persistence will influence the prospect and increase your chances of closing the deal. At some point, the prospect will begin to believe that your persistence comes from your conviction that your product or service is the best available. Often, persistence is the key that tips the scale in your direction. Frequently, it is the most persistent salesperson who prevails.

I find it effective to reinforce several key benefits when recontacting prospects. I might start out by saying, "As you remember, Mr. Prospect," (then I bullet point a few benefits worded in a different way than detailed in the original presentation). Never ask them what their decision is. Often, I will make another appointment for the purpose of showing them something new. Try to get eyeball to eyeball with the prospect. I can't remember the last time I closed a sale on the telephone.

Most sales are finalized on the fourth or fifth closing attempt. The average salesperson *might* make one or two closing runs and quit. Could this be the reason we have so many low achievers in the selling profession?

I became a territory sales rep selling mechanical, rotary calculators and adding machines for SCM Corporation early in my career. My territory consisted of a few square blocks in the financial district of San Francisco. I had set my goal to become the number one sales rep of eighty in the western region. The branch manager suggested that I was shooting too high for the first year and recommended that I go for top spot in our branch office.

I couldn't wait to hit the street, but first I had to endure a week of product training in the office. It was raining hard the following Monday as I left the office with my left arm around a demonstrator calculator. I stepped out between two parked cars to cross Market Street and a speeding car hit my left hand against the calculator, severing my thumb and crushing the back of my hand. My thumb was dangling by a thread of skin.

The branch manager drove me to the emergency hospital. A doctor who could have been Paul Newman's twin sewed the thumb back on and sent me home with orders to cool it for the rest of week.

I went to work the next day with my left arm in a sling, a cast on my thumb, my hand stitched and fully bandaged. I decided to show up at the home office of a major marine shipping company and try to get face to face with the manager to demonstrate our latest calculator. My boss and his boss each had had my territory previously, and had not been able to meet with this guy over a period of about three years. I had called his office several times without making contact with him. His calls were heavily screened.

It was eight blocks to his office from where was I parked. I was totally drenched by the heavy rain, and my hand was leaking blood through the bandages. The elevator stopped at a large open area facing a semi-circular, enclosed reception desk. The receptionist, her back to me, was busy sorting some papers on the credenza. I quickly moved to the front of the desk, dropped down to my knees and started crawling quietly along the front of the desk in the direction of a hallway.

Once out of sight of the receptionist, I stood up, breathed a sigh of relief, and started down a long hallway of offices with old-fashioned wooden doors and smoky glass windows that you couldn't see through.

I recognized the elusive manager's name on the door at the end of the hallway. I opened the door and walked into a small office. Sitting at his desk was a large, ruddy-faced Irishman. There was less than four feet between us. I'll never forget the surprised look on his face. I stood perfectly still as I looked at him with a smile on my face. He finished his conversation, and as he was putting the phone down, he asked in a loud, demanding tone, "Who are you?"

I replied, "Gerry Shaltz, I'm with SCM Corporation."

He queried, "What do you want?"

I answered, "I want to know how to get an appointment with you to show you our latest calculator."

His voice became louder as he said, "You have to call for an appointment. You don't go walking into my office unannounced."

I replied, "That's why I'm here, we've been trying to make an appointment with you for over three years and haven't figured out how to connect with you." I just stood there motionless as we stared at each other.

Suddenly, I could see the madness of the unfolding episode through his eyes. Here's this crazy, strange guy barging into his office in a wet crumpled suit, with blood dripping from a bandaged hand in a sling, asking how to make an appointment.

He said in a gruff, threatening tone, "Get out!" I was frozen in place and didn't know what to do or say next. I didn't want to leave until he agreed to a demonstration.

Next, he stood up, pointing his finger toward the door and yelled, "I said, get out or I'll have you thrown out!"

"I'll leave", I said. "But, I must tell you one thing first." I continued, "Yesterday I was on my way here and I got hit by a car. My thumb got knocked off, sewed back on and I came back today against doctor's orders. I parked eight blocks from here. I crawled in front of your receptionist's desk so she wouldn't see me. Now, I'm willing to walk eight more blocks back to my car to get a forty-pound calculator with one arm and walk back another eight blocks to demonstrate it for you and then go back to my car when we're done. That's thirty-two blocks and it's raining like hell out there. I don't know how you got to where you are. But what I'm trying to do here has got to mean something to you." I knew enough to shut up at that point.

It felt like an eternity before he spoke. He sat down and said, "Go get the damn machine." I sold him three calculators and proceeded to close four more throughout the remainder of the day. The orders looked like a chicken had scratched them. I'm left handed. So, I had to use my right hand to fill out the purchase order.

I finished the year as top rep in the western region at 456% of quota. That was good enough for the number one spot in the nation, out of hundreds of sales reps. Then I quit. I hated the job.

(8) They Set Goals

"People with goals that are clear, and written down, accomplish far more in a shorter period of time than people without them could ever imagine."
                                              Brian Tracy

Here are the five basic instructions for setting and achieving goals:

(a) **Set your own goals** – Don't allow others to do it for you. Also, if you decide to make your sales quota a goal, reset it somewhat higher than that which has been assigned. Then break down the weekly and daily activities necessary to meet that particular goal.

(b) <u>**Write down your goals**</u> – Something amazing happens when you write down your goals. I can't completely understand why, but it does. It begins in your brain, travels down your arm to your fingertips to the pen and onto the paper. You will have your goals written in your own words where you see them take form on paper. This starts the mental imprinting.

(c) <u>**Discuss your goals**</u> – Do this with someone you trust and respect. This is a critical step to goal setting. The goal then takes the form of a commitment, a mutual expectation that is permanently imprinted on your brain.

(d) <u>**Keep a copy of your goals handy, and read them frequently.**</u>

(e) <u>**Here is the one foundational rule for everyone in sales. It's an absolute, and should not be taken lightly by anyone who wants to succeed.**</u>

"You can't control results, but you can control activity. If the activity is right, the results are automatic."

J Douglas Edwards

All you need to do is to define the activity. You must break it down into small increments. Here's an example of the process: "Your goal is to sell a certain number of widgets or programs (or whatever it is) on a weekly basis." Then, determine how many presentations you need to give in order to arrive at that weekly number of successful outcomes.

Your plan will begin to take shape. You'll know what you must do each day in order to achieve your weekly, monthly, and yearly performance goals.

I don't know of a more effective way of ensuring success. Break down the activity into daily action steps. This is a powerful lesson. Learning it and doing it is the one thing that quickly elevated my career by several levels. I got noticed. Wonderful opportunities started coming my way.

"Goals are dreams with deadlines."

Unknown

(9) They Engender Trust

"Technique and technology are important, but adding trust is the issue of the decade."

Tom Peters

Trust is everything. You *earn* trust by being trustworthy in all your dealings with the prospect throughout the entire selling process and after. **Partner with your prospects.** Work together with them to address their needs. Be a trusted consultant. Consultative selling is powerful. It will automatically give you a competitive advantage. Consultative selling means that you and the prospect are on the same side of the table. You are providing valuable, expert information and advice while doing and suggesting the things that are in the prospect's best interest. Do not make the mistake of representing any product that is not in the best interest of your prospect or client.

While you emphasize the benefits of your product, respect your prospect's intelligence. They know that no product or service is perfect. Be forthcoming about shortcomings and point out how the advantages outweigh the disadvantages. Provide the pertinent information. Know your product and/or service. Most importantly, emphasize the benefits of what you offer and how those benefits will resolve your prospect's problems and/or needs. I've found that when a person wants something badly enough, he/she will create rationale to justify acquiring it.

The "want" becomes "I need". Always keep in mind the radio station that everyone tunes into, **WIIFM; WHAT'S IN IT FOR ME**.

(10) They Are Driven to be The Best

"People become really quite remarkable when they start thinking that they can do things. When they believe in themselves, they have the first secret of success."

Norman Vincent Peale
Author, The Power of Positive Thinking

Competition is the key to excellence. Top salespeople have a strong need to compete. They want to be in the game, and they want to win. They expect to win. If you are not competitive, you will probably fail in the selling business. We often compete with others to provide better services and products. Thus, our clients win, we win, and our company wins. Competition sharpens skills. Some are fierce competitors because they need the money or keep score by it. Others need to win because it's how they get in touch with themselves. It's OK to compete for your own sense of self-satisfaction, to be number one, or simply to be noticed.

(11) They Have A "Do It Anyway" Mentality
(They Do What Others Are Too Afraid, or Unwilling To Do for Success.)

"Professionals can do their best work even when they don't feel like it."
Alistair Cook

Do It Anyway! The real pros do the things that are hard to do, things that they may find unpleasant but necessary in order to be successful, even if that means sitting down in a chair and facing the difficult task of making telephone cold calls for appointments. The telephone can be a major source of rejection. Remember, Babe Ruth set the record for home runs while setting a record for strikeouts. He kept swinging anyway.

(12) They Actively Work to Improve Their Skills

Champion salespeople maintain an active and ongoing interest in improving their skills. They read books and attend seminars on selling. They understand the need for growth. You can't tread water in sales or in business. You are either growing or dying. I believe that the same holds true for physical and spiritual growth. It's like being out in the middle of a lake. One can tread water for only so long before fatigue sets in and one begins to drown. Each of us is either growing or dying. Treading water is an illusion.

I sat next to the CEO of a Fortune 500 company several years ago while flying back to Los Angeles from New York. Because we had both come up through the sales ranks, our conversation was focused on the importance of selling. He asked me this question, "How many years of sales experience do you have?"

"About forty years", I replied.

Then he asked, "Are you sure it's forty years of experience?"

I simply replied, "Yes, I'm sure."

He added, "Are you sure that you have forty years of experience, or just one year repeated forty times?"

A short pause in the conversation ensued. I smiled at him knowingly and said, "Yes, I'm sure." Point well taken: **grow or perish.**

(13) They Sell Value – Not Price

"Fewer consumers than you might imagine buy based on price. Yet, far too many salespeople sell price to most consumers."

Price becomes far less important when people understand the value. It takes no talent to lower your price. However, it takes product knowledge, talent, and perseverance to convince the prospect of the superior value of investing in your product and in you. Stress value by educating the prospect, spending time with him/her, and de-emphasizing the price.

(14) They Anticipate Objections

As you become more experienced at giving presentations, you will learn what kind of objections/ questions you are most likely to hear. You need to build your presentation with that in mind. Anticipate the objections and answer them as part of the presentation. Thus, you will decrease the number of objections you receive.

I suggest that you pose the following three questions prior to starting your presentation:

(a) "Is there anyone else that you would like to include in this meeting?" Often, this question eliminates the need to repeat the presentation to others. It also helps to identify and gather the decision-making body prior to giving your presentation.

(b) "Every business makes decisions differently. How does your company go about making decisions like this?"

(c) When competing against an incumbent product or service, ask these two questions, the answers to which will guide you to the prospect's "hot buttons": "What do you like about your present service or product?" "If you could improve it, what would you do?"

NOTE: Keep control of your materials. I've witnessed salespeople handing the prospect a brochure prior to the conclusion of the presentation. They must then give a presentation to a prospect who is reading the brochure at the same time. **Keep your materials under your control.** Pass them out when it makes sense to do so, and when it's in the best interest of the sale.

(15) They Utilize Tie Downs

Build "Yes" momentum with tie downs. What is a "tie down"? The "tie down" label and concept was taken from, *How to Master the Art of Selling* by Tom Hopkins. It's a semi-rhetorical question you ask the prospect immediately after presenting a key benefit in order to develop "yes momentum." Some examples would be, "Can you see the benefit in this?" "Is this what you had in mind?" "Isn't this great?" The more "Yes" momentum you can build; the easier it is for the prospect to say "Yes" to the final question. Can you think of more good examples of "tie down" questions? You can never have too many.

(16) They Say Exactly What They Mean
(They Are Great Communicators)

"They say exactly what they mean without embellishment. Time is currency!"
Richard Weylman
"Be an Effective Communicator"
Marketing Tip 11/27/07

"When we communicate verbally with others, either in a conversation or in a presentation, our usual goal is for people to understand what we are trying to say. In order to accomplish this, we should remember the acronym KISS (Keep It Short and Simple). When we talk to others, we assume they will understand us. We know what we are trying to say, so obviously our message will get through. Right? Not necessarily. People bring their own attitudes, opinions, emotions, and experiences to an encounter, and this often clouds their perception of our message.

It's Up To Us

"When we speak, only approximately 10% of the words we use get through to others. Spoken words are unlike written words where a person can go over a passage several times to ensure understanding. It is our responsibility to make sure our message gets across to our audience. Therefore, if we want our message to be understood, we must be careful of the words we use.

"When we communicate, we need to put ourselves in our listeners' shoes. Put yourself on the other side of the table. How would the message sound if you were not fully versed in the topic? Would you understand the message or would its meaning be lost on you?

Choose Words Carefully

"Very often when individuals are extremely well versed in a particular field, they might have a tendency to use industry jargon in conversations or presentations. While this may be comfortable for the expert, it often causes confusion on the part of the listener. If, for example, you are discussing computers, you might be talking about bits, bytes, CPUs, and controllers. However, unless you are talking to someone who is equally well informed about computers, that person will have no idea what you are saying.

"Other people believe they are like Charles Dickens who got paid for every word he used to tell his stories. They think the more words they use to describe an idea or concept, the more effective they will be in getting their message across to others. It might have been advantageous for Dickens to use many words to express his ideas, but for most of us, keeping our words succinct and to the point allows our listeners to understand what we mean.

"Still other people believe using long or difficult words will impress their audiences. While using an extended vocabulary is impressive, if someone needs a dictionary to decipher your meaning, your message will be lost. If people have to work hard at trying to understand what you are saying, they probably will not put forth the effort.

"Once again, our objective in verbal communication is to have our message understood by our listeners. In order to achieve our goal, when speaking with others, always remember to…KISS."

By Della Manechella,
Personal Peak Performance Unlimited.
E-mail: Della@dellamanchella.com

Word Smart

Words influence the outcome of a sales presentation. Words can change minds. People judge you by the words you use. Become *word smart.*

Words and Phrases People Like

I couldn't agree more.
Protected
Fully protected
Safe
I respect that.
I understand how you feel.
Statistics show...
Makes sense.
Time is money.
Options
Fair enough?
Risk management
Right on.
You're ahead of the game.
It pays for itself.
Invest
Investment (instead of buy)
I need your approval.
Can I make a suggestion, please?
I appreciate that.

Please work with me on this.

Seems logical.

It's easy.

You can do it.

It works.

You're the boss.

Thank you.

Let me tell you a story.

Guarantee

Please allow me.

(17) They Differentiate Themselves From the Competition

The chances are rare that the competition has achieved all the objectives outlined in **Lesson 2.** Therefore, the first step in differentiating yourself is by making certain that you **achieve those objectives.**

Be personable. Build rapport. Remember that approximately fifty percent of prospects choose to do business with people they like.

Find their "hot buttons" in advance of your presentation. A well-planned, logical presentation that meets their needs and connects with their "hot buttons" can give you a distinct advantage.

When prospects **perceive you as an expert** in your field by way of the pertinent knowledge you impart and the unique and useful ideas you offer, your chances of earning their business are greatly increased.

You differentiate yourself from the competition by your **successful appearance and professional demeanor.**

Preparation will reduce fear and increase your confidence. Understanding your prospect's needs better than your competition is a core differentiator.

Asking better questions will make a difference and impress the prospect.

Following up with e-mail, hand-written notes, and letters summarizing the meeting and the benefits of your product after each contact with the prospect will further project a professional image.

You differentiate yourself by maintaining a **higher level of enthusiasm** for your product.

You impress your prospects by **showing more interest in their business** than the other guy does.

Summary

Be personable
Find hot buttons
Be the expert
Present a successful appearance
Present a professional demeanor
Be prepared
Ask better questions
Follow up

Maintain high enthusiasm
Show more interest

(18) They Ask for the Order

> *"If you can't close, you can't sell."*
> *(from the magazine,* Smart Business Ideas)

Michael Bloomberg, mayor of New York City, was quoted as saying the following: "I can't remember who told me this, but I certainly didn't grow up knowing it, so I must have gotten this advice at Salomon Brothers in the 1970's. The advice was, first, always ask for the order, and second, when the customer says yes, stop talking. The worst advice that people can take is to react before they've had a chance to think. I think we all say things and wish we hadn't said them. Ready, shoot, aim is not the smartest policy." In other words, ask for the order and shut up! Do not break the silence."

By Michael Bloomberg,
Fortune 500, "Best advice I ever got."

(See Lesson 6, Closing the Sale)

Objectives and Components of a Winning Presentation

I decided early in my career that it would be valuable to identify the load-bearing pillars that support a sale. One might ask, "Why do we need more than one objective when it's obvious that closing the sale is the objective?"

The answer is best explained by asking these questions:

Against what criteria can you evaluate your performance after the presentation, whether or not you've earned the business? How can you prepare and deliver a presentation if you don't know the fundamental pieces that support the sale?

I've seen newcomers fail because they got lucky and closed a deal early on with a sub par presentation. They continued to utilize the same poor presentation and failed. How could they know the reason that they failed?

This lesson contains a proven checklist of the critical objectives that must be achieved during the sales presentation to ensure the best possible outcome. I created the list so that I would have a running track of the most important things I had to achieve during the presentation in order to earn the business. These objectives are the result of much trial and error over thousands of presentations.

Once completed, the list allowed me to review the objectives prior to a presentation and evaluate my performance post presentation. It enabled me to make adjustments during the presentation when circumstances moved me out of sequence or forced me to compromise an objective. These are tools that make it possible to adapt "on the fly" while giving the presentation.

For example, building rapport prior to your presentation is one of the pillars that you may not be able to achieve each time. Recognizing that rapport is an objective; you would seek out opportunities during the presentation to develop or further enhance that objective.

Each pillar is so critical to building a strong foundation that it's impossible for me to tell you which one is most important. I cannot tell you which objective I would eliminate and still earn the business. Achieving each objective is critical for a successful outcome. Miss four or five and you might as well forget it.

(1) Know the audience

(2) Get their names straight

(3) Include some humor

(4) Have it flow in a logical sequence

(5) Utilize visuals

(6) Minimize technical aspects

(7) Maximize benefits

(8) Amplify the problem

(9) Present the solution

(10) Consider how it will make people feel

(11) Include common objections

(12) Instill urgency to buy now

(13) Keep it short and simple

(14) Ask for the order

(15) No close. … why not?

(1) **Know the audience** – A sales presentation given to an audience without the presence of the decision-making body is usually a waste of time, unless you need the practice. Some companies purchase by committee. In such cases, you must work your way, step by step to the final decision maker. However, the higher up you can start, the better off you are.

Present to the decision-making body. I will not subordinate my presentation to anyone who doesn't have the authority to make the decision or, at the very least, to lead me to the decision maker. I would rather use the time, effort, and resources to find a more viable setting to ply my trade. I would place the contact in my "later" file and get back to it if and when I have nothing better to do. While it's critical to always try to give the presentation to the decision-making body, sometimes you are forced to comply with company policy and work your way up, step by step through committees. If the reward is big enough, I may comply, especially if I'm wearing the boss's badge of authority. Make sure you close at each level as you make your way up to the final agreement. The close at each level is an endorsement from the present group and an appointment set with the next one, with the agenda (ground rules) agreed upon in advance.

I was trained by a man who managed a group of us as we handled sales to elementary schools. He was the top seller and group leader. We left at 5:00 am and drove from San Francisco to Bakersfield. Upon arriving, the superintendent said, "Something has come up and I can't meet with you. I would like you to meet with my curriculum coordinator, John. He is in the building just around the corner."

Most people would have gotten in their car and rushed over to meet John. Not Bob. He asked, "Can John make the decision to go ahead?"

The superintendent said, "Yes."

Bob asked, "Does he make the decision to recommend this to you or does he make the final decision to implement programs district wide?"

"No, he makes the final decision. He's been with me for many years and I trust him."

At that point, 99 out of 100 salespeople would have run down the street to meet with John. Not Bob. "Can John approve the agreement?" Notice that he didn't say "sign".

"Yes," the superintendent said.

By that time, I was up and ready to go see old John. Not Bob. He said, "Does he know he can?"

"Uh, I don't know", the superintendent replied.

Then, Bob asked, "Would you do me a big favor?"

"Sure."

"Would you tell John that he can?"

The superintendent called John and told him that there were a couple of fellows coming to see him with an enrichment reading package and if he liked it, he could sign the agreement. Then, we left. Guess what? We got the deal. That's called "selling smart."

That was a lesson I will never forget. The ROI (Return-On-Investments) for callbacks was really low in that business, so we didn't make callbacks. We had to pay all of our own expenses, including hotel and car. We had our own money invested in every presentation. So, it was important to take steps that were the most likely to bring success. This job was a true boot camp for me.

(2) **Get their names straight** – Draw a simple seating chart including the names and positions of each attendee, especially if there are three or more people in the room for your presentation. This will avoid the potential embarrassment of addressing people by the wrong name. Having to ask, "What's your name again?" is, at the very least, a distraction.

Also, make certain that you can pronounce their names correctly. Address people by their first names if appropriate.

(3) **Include some humor** – Humor is a great ice-breaker. It relieves tension. Tickle their fancy and you'll own them. Avoid off-color or politically incorrect jokes. Personal experiences are best. You will be perceived as more "human" when you can laugh at yourself.

(4) <u>**Have it flow in a logical sequence**</u> – It needs to flow, step by step, in a logical sequence toward the close. Keep the close in mind at all times.

(5) <u>**Utilize visuals**</u> – Most people are more visual than auditory. They remember two or three times more of what they see than what they hear. Create a presentation that will deliver a high visual impact.

(6) <u>**Minimize technical aspects**</u> – Even if you're presenting to a group of engineers, it's far more effective to emphasize the benefits of your product. Too much detail can be boring.

(7) <u>**Maximize benefits**</u> – Show prospects how your product saves them money, how it increases company efficiency, how it gives them a competitive edge in the market place. Demonstrate what your product can do for them.

(8) <u>**Magnify the problem**</u> – Selling is a problem/ solution relationship. You wouldn't be there if they didn't have a problem or a challenge that you can effectively solve.

(9) <u>**Present the solution**</u> – They need more efficiency, innovation; they need to save money; and they need to do something better or faster. They are looking for an advantage over their competitors. Today's marketplace is crowded, competitive, and smart. People are interested in how you meet their needs. They will be most impressed by your research and how well you understand their needs. The close is influenced by what happens throughout the presentation.

(10) <u>**Consider how it will make people Feel**</u> – No purchasing decisions are made without feelings. We are feeling creatures, and it's up to you to determine how your presentation will make your prospect(s) feel.

(11) <u>**Anticipate objections**</u> – As your experience at giving presentations grows, you will begin to know what kind of objections you are most likely to hear. Build your presentation with that in mind. Anticipate the objections, and answer them as part of the presentation before they occur.

(l) <u>**Instill urgency to buy now**</u> – It's not hard to do. Give them reasons to invest now. Perhaps they'll lose money by waiting. Maybe they'll forego potential profits. Perhaps there's going to be a price increase soon. Or perhaps the demand is so great that your company is back ordered. Any number of reasons might come into play. Use your imagination to create urgency. The goal is to leave with a finalized agreement whenever possible.

(13) **Keep it short and simple** – Remember that people need to be able to consider whether the solution is feasible. Don't over complicate your presentation with unnecessary details. This can create confusion and doubt. Help them form simple, mental pictures of the result of owning the product, the benefits of investing in your product and in you. Ask questions. Keep the prospect talking. Listen more than you talk.

(14) **Ask for the order** – I guarantee that there is an astonishing percentage of salespeople who never ask for the order. They are either afraid to ask, or don't know how to ask. Many just keep talking until they are beyond the opportunity to close. Finally, bored, the prospect asks them to leave some literature to get rid of them. Some don't have a clue when to close. Others think they're in front of the prospect to simply disseminate

information. A trained chimp could do that. Keep in mind that the entire sales process is one big close.

(15) <u>No close. ... why not?</u> – I have a huge investment in every presentation I make. This includes every selling failure I've ever experienced, countless hours of networking and setting appointments, research, thousands of miles driving to and from appointments, numerous selling seminars and lectures attended, books on selling read, thousands of hours preparing and practicing presentations. Each and every presentation represents a ton of resources, preparation, and sweat equity.

As a professional, I feel that this gives me the absolute right to ask the prospect to buy or to know why he is not buying. Additionally, I know that I will not have a chance to close the sale if I don't find out why the prospect is resisting. There is no better time than right then, while everything is fresh in his/her mind, to pose the question: "Mr. Prospect, can I ask you a simple question? As a courtesy to both of us, please tell me why you aren't going along with this?" I've closed many sales by simply asking that very question.

Once you know why the prospect is resisting, you have a perfect opportunity to address the issue and go for the close. Ask: "Is there anything else causing you to hesitate?"

Prospect: "No."

Salesperson: "Then, if we can satisfy this concern, can we earn the right to your business?"

Defining Moments that Threaten the Sales Process

There's a crucial moment that frequently occurs during the sales process. It's a defining moment that goes completely over the head of many salespeople. The best analogy I can offer is this. You're driving down the highway at 70 mph. Visibility is good and traffic is light. Suddenly, a big brick wall appears directly in front of your car. Your options are limited. You can hit the brakes and maneuver around the wall or hesitate and suffer the consequences. Unfortunately, thousands of potential sales end in wrecks because the salesperson doesn't see the wall or know how to maneuver around it.

While defining moments can pop up at anytime during the sales process, the major defining moment occurs when the decision maker tries to turn you over to a subordinate, a recommender, before you've made the presentation. Sometimes this happens when you're on the phone trying to make the appointment to give the presentation. Often, it happens in the first few moments of meeting with the decision maker.

Most salespeople comply in the name of pleasing the prospect and keeping the opportunity alive. At that very moment, the chances for a successful outcome have been severely compromised. The salesperson may never understand why the deal was lost. After all, he/she met with the highest authority and did what was requested. "It wasn't my fault. What else could I have done?"

It may not be in the recommender's best interest to support your effort for any number of reasons. He/she may feel that there's nothing to gain by doing so. Perhaps the recommender feels that there's personal risk attached if the product or service recommended to the boss doesn't deliver. The recommender may have a "personal" relationship with the incumbent provider. The prevailing decision to play it safe, say little or nothing to the boss is a common outcome.

Nothing ventured, nothing lost. One can be assured that most people will do whatever they believe is in the best interest of protecting their income.

Once you allow yourself to be moved over, the best you can hope for is that the recommender will present your product to the decision maker. In which case, you're betting that he/she can present it as well as you can after hearing the presentation one time.

Once you are in the hands of the recommender, you have most likely been relegated to a marathon of follow-up, phone messages, excuses, unrecoverable time, and sheer frustration. You may close the odd one. I cannot tell you which one. However, I guarantee you that your long-term rewards will be minimal, at best.

I prefer to declare myself with the decision maker in those few moments and convince him/her to attend the meeting with me. I do this even when it seems like I might have to walk out. Worst case, it would free me up to spend my time contacting other prospects. Not such a bad choice. It took me a number of years to learn this lesson. I wasted a great deal of unrecoverable time chasing subordinates in the name of goodwill. It cost me plenty. Some never learn the lesson.

Every top salesperson understands that there comes a time when taking a stand to protect one's interest and investment is necessary. They understand what it takes to achieve the objective. They understand the value of their time. They know the odds. They avoid situations that are proven to steal their time. They fully realize that lost time is unrecoverable.

Whenever a decision maker attempts to turn you over to a subordinate to give your presentation **do not comply immediately.**

Say the following: "I would very much like to meet with John. However, since everyone thinks differently, you may have questions that John doesn't ask. Also, you probably view this business from a different perspective than others. For that reason, I would appreciate the opportunity to present this product/ service to you and John. It only takes about fifteen minutes. If today isn't convenient for you, let's check our calendars and reschedule a better time; fair enough?"

Prospect: "Well, I'd still like John to see this first."

You: "I can certainly appreciate the fact that you rely on John's opinion in these matters. This decision is so important to an organization that we've been meeting with presidents even in the largest companies. Let's go ahead and reschedule with both of you for a better time. Do you have your calendar handy?"

Prospect: "I want you to meet with John first"

You: "If John is in favor of our program, can he make the commitment to proceed?"

Prospect: "No"

You: "What would be the next step, assuming he wants to proceed?"

Prospect: "He would recommend it to me and I would make the decision."

You: "Since John doesn't know who I am, could you introduce us and let him know that you want him to meet with me?"

Prospect: "I'm really too busy to do that right now."

You: "Can I tell him that you want us to meet?"

Prospect: "Certainly."

You: "Thank you. I'll follow up with you as a courtesy and let you know the date and time of my meeting with John. Perhaps you can sit in. I'm confident that you'll feel the time spent is worthwhile. Thanks again for your interest. I'm looking forward to seeing you again."

Let's review what has happened in this scenario:

(1)   Three attempts were made to keep the decision maker in the meeting.

(2)   One attempt was made to clarify if the recommender could make the decision to proceed.

(3)   A mutual understanding was reached with the decision maker about how the decision will be made.

(4)   An attempt to have the boss tell the recommender to meet with you was made, which would be a valuable, implied endorsement.

(5) Permission was obtained to use the president's name and authority.

(6) A gentle reminder was issued to the president that we'll still try to get him to attend the meeting.

(7) Note that after each attempt, the decision maker had an opportunity to say "Yes" to your request, and many do so somewhere down the line of attempts. After your first two attempts to keep him/her in the meeting, it's not uncommon to hear the decision maker agree, "OK, how long did you say this is going to take?"

However, if you decide to comply and meet with the recommender, you are far better off when you wear the boss's badge of authority. You will have the subordinate's full attention for no other reason than that he/she knows that you and the boss have discussed the presentation. Remember to stop by the boss's office when you're leaving. If possible, you can invite the decision maker to join the presentation when you meet with the recommender.

It won't come as a surprise to the boss to hear from you again to announce that you are about to meet with the recommender. This is an opportunity to extend another invitation to him/her to join the meeting. You have also earned the right to update the boss post meeting in the event he/she has chosen not to attend. Each encounter is an appropriate closing opportunity.

To summarize, it's important to remember that there is no better audience for a presentation than the one with the highest authority, the person who can make the final decision to proceed. Meeting with the recommender is a very distant second, especially if you're not wearing the boss's badge of authority.

Prospects often request literature prior to the presentation: "I'm kind of busy. Can you leave some information?" Delivering printed material to a prospect prior to a presentation is weak; it's an excuse for not selling. Hoping that the prospect will read the information and contact you for a presentation is futile. It may happen occasionally. However, don't expect to earn a professional income by handing out brochures. Most are placed directly into the round file. Reschedule the appointment if the prospect is too busy to go forward with the meeting. Top salespeople find a way to meet face to face with decision makers.

LESSON 4

Telephone Power
(Setting Appointments)

I hired a salesman whose references checked out very well. He was the top salesperson at his previous company. The guy failed. He was so afraid of the telephone that his hands would start shaking the instant that he picked up the handset. Cold calling on the telephone can be a prime catalyst that amplifies the fear of rejection. He eventually sought the help of a psychiatrist, but still couldn't overcome his fear of the phone. The reason he'd been so successful at his previous job was that the company had a telemarketing department that set up the appointments for the salespeople. Once he was with a prospect, face to face, he was great.

Setting Appointments: This is a subject that I spend a great deal of time on in my seminars. I have participants drill for skill by role-playing.

Setting the Appointment:

(1) Write down the full name and title of the person with whom you wish to speak.

(2) Have a written phone script in front of you including words to deal with the gatekeeper.

(3) Understand that the objective is to schedule an appointment. The appointment is the close. Many salespeople give too much information in their quest to make the appointment. Because of this, the prospect feels pressured. Only give enough information to get the appointment.

(4) Be friendly and smile. People can feel your smile over the phone. I was conducting "phone power" training with a group of salespeople when I noticed that most of them were so tense that they weren't smiling. The prospect can sense tension. I asked the students to use a small mirror, write the word, "smile" across it with lipstick and place the mirror so they could look directly into it.

(5) Use the person's name frequently. People like to hear their name. It is music to their ears.

(6) Wait a few seconds (a beat or two) after they answer the phone.

(7) Ask, "Did I catch you at a good time?" (Take the curse off of the call).

(8) Be brief.

(9) When questioned, answer briefly and close for the appointment. Ask, "Do you have your calendar handy?"

(10) Repeat the appointment time and date. (Make sure they write down your name, the time, and date of the appointment), and have them repeat it back to you.

Dealing with the Gatekeeper

Most telephone receptionists screen calls, particularly for upper, level management. Our objective is to defuse the receptionist from controlling the screening mode, thus increasing one's percentages for making live telephone contacts with the decision makers. Always give the receptionist your full name. Why? Asking the caller his/her name is the first step in the screening process. Once you've given your name, it's not likely you will be asked for it. "Hi, this is Gerry Shaltz. I'm calling for Bob Johnson. Is he there?" Common courtesy dictates an answer to the question "Is he there?" If Bob Johnson is there, the receptionist is not likely to reply, "no." The receptionist might ask, "What is the purpose of your call?" or, "What is the name of your company?" My stock reply is: "I really need to speak with Bob personally. Could you please put him on? Thank you." It's not advisable to leave a message unless the prospect knows you and you believe he/she would return your call. Try several more times to catch the prospect in, so you can have a live, one-on-one conversation. Giving your name immediately to the receptionist and asking if prospect is there will not result in a connection to the prospect each and every time. However, it will help to neutralize the screening advantage and result in a higher percentage of successful, live contacts with prospects. In the event there are a very small number of prospects for your product or service, you may find it advisable to leave a message earlier rather than later.

Intruder

Decision makers are not sitting around all day hoping for phone calls from salespeople they don't know. Such calls are an intrusion on a busy person. As an intruder, one has a very short window of time (seconds) to capture the prospect's attention.

Stimulate interest early in the call by:

(1)   Your professional and courteous manner

(2)   Using the prospect's name more than once

(3)   Being brief and direct

(4)   Creating the expectation of **gain,** sparking their **curiosity,** and/or amplifying the **fear** of loss (See the curiosity, gain and fear objectives on the following page.)

Example: "Mr./Ms. Johnson, in the next few moments I'm going to offer you an opportunity to learn about a system that has been consistently successful at increasing the bottom line profits of hundreds of companies." This sentence contains the promise of reward and the element of curiosity. The prospect may have felt the fear of loss since he/she didn't end the call. All you have to do is introduce any one of the three items. Obviously, the more the better. One is well advised to utilize this tactic in every presentation as an effective tool for holding the prospect's attention.

This is one of the most powerful cannons in the selling field. Unless you can accomplish one of the three following objectives in the first few seconds of the telephone call, you probably won't get the appointment. Likewise, unless you accomplish one of these three things early in the sales presentation, you probably won't close the deal.

(1)   Spark their **curiosity.** If people are curious, they are going to pay attention.

(2)   The **promise of gain** is guaranteed to hold someone's attention.

(3)   Amplify the **fear of loss.** If people are afraid that they could lose something, they are going to pay attention. You need to keep these three rules in the forefront of all your selling.

Telephone Appointment Setting Script

Three Steps:

(1)   Take the curse (intrusion) off the call.

(2)   Create interest.

(3)   Sell the appointment.

You: Mr. Johnson?

Mr. J.: Yes?

You: Pause (2–3 beats) – Thank you!

Mr. Johnson, it will just take a moment for me to explain why I'm calling. Is it convenient to talk now?

Mr. J.: Well, go ahead.

You: Thank you!

Mr. Johnson, I represent _____, my name is _____.

Our company has been doing some work with businesses very similar to yours throughout the country and has developed ideas and techniques that have proven highly successful at generating more bottom-line/incremental income for these companies.

The ideas and techniques we've developed could be very valuable information for you.

It will take just a few minutes to cover these ideas with you, and if I'm there any longer than that it will be because you've asked me to stay.

Could I see you tomorrow morning at 9:15 or would Thursday morning be better?

Mr. J.: What's this all about?

You: It's about sharing some proven ideas and techniques to increase your profits that other companies, like yours, are utilizing.

Would 9:15 tomorrow work for you, Mr. Johnson? (I always suggest meeting times in quarter hour increments, as it gives the impression that you are tightly booked)

Button down the Appointment

Note: When training for phone power skills, separate the two individuals who are engaged in the role-playing so that they cannot see each other. This simulates a real telephone call, eliminates face-to-face contact, and facial and body language clues. Also, it sharpens one's auditory sensitivity to the posture of the prospect's voice.

(Credit for much of this phone script to: *KISS: Keep it Simple Salesman - Selling Techniques That Really Work,* by Earl Nightingale and Fred Herman.)

LESSON 5

Handling Objections

Building rapport and handling objections are the two most fluid parts of the selling process, and equally important. Dealing effectively with objections starts with your mind set. I was supervised by an individual who never heard an objection. He heard only questions and always responded in kind. His demeanor remained constant. He would start his response by saying, "In other words, Mr./Ms. _____, you're **asking** if _____." This mindset is a powerful tool that helps you to maintain a professional image and to stay calm.

The fundamental rule for handling an objection is to hear it as a question and/or a request for

more information. It's important to remember that the prospect may simply be curious about something that is obviously not material to the sale. This wouldn't be a question that indicates resistance; rather it's informational in nature, not an objection. In which case, the only rule that applies is to answer the question and move on.

Pause for a moment. There's a space between stimulus and response where you must sit when you hear an objection; reflect silently for a few moments, then respond. This gives you time to formulate your response and slow the pace. It also demonstrates respect when you take a few quiet moments to consider the prospect's objection.

Repeat the objection to the prospect in your own words and ask if it's what he/she meant. This is a communication checkpoint. The worst thing you can do during the exchange is to provide an answer that has nothing to do with the objection. I've done it. It's embarrassing and an automatic turn off. You may not have correctly understood what the prospect was saying. By translating it into your own words, it allows you to make certain that the communication is clear. I've had prospects say, "That isn't what I meant", and proceed to restate and clarify. Once you understand it, you can deal with it. **Clarity is the key.**

Asking the prospect to explain why he/she feels that way is an effective tool. "Do you mind telling me why you feel that way?" This question not only buys you additional time, it's an opportunity to learn more about what is important to the prospect. Sometimes, while trying to explain, the prospect sees that the objection is unimportant or doesn't make sense, and she/he gives up. Oddly enough, I've had prospects say things like, "Aw forget about it, it's really not that important anyway." The objection was so inane that the prospect could not explain it.

The three F's are a well known and frequently used tactic. It goes like this: "I understand how you **feel.** Many of our clients **felt** the same way. However, they **found** that after seeing the significant results in

_____ .

This response may seem trite at first blush. However, its power lies within the effect it creates. "I understand how you **feel**," shows empathy that can put the prospect at ease. "Many of our clients **felt** the same way," assures them that they're not alone; not the only ones who felt that way. "However, they **found**," is a key transitional phrase to start the response to the objection. I utilize this when I can't think of a better response, which is often.

Many people are familiar with the 3 Fs. They may even smile knowingly when you use it. Don't let this inhibit you. This is a tactic that works. *Even when they know what you're doing, they can't resist.* Return the smile briefly and keep going forward.

Look for opportunities to close on an objection. Many talented salespeople firmly believe that one should attempt to close on every resistance. Ask the prospect if the objection is the only thing standing in the way of doing business. "That aside, is there anything else preventing us from earning your business?" Once you've reduced the exchange to a final objection, say this: "If we can satisfy you on that point, can we go forward and finalize this?" If the prospect says "Yes," upon proof, you've closed the deal.

You have the option of ignoring an objection. I worked with an outstanding salesman who wouldn't respond to an objection unless he heard it twice. It made me nervous the first couple of times that I watched him give his presentation. He would not so much as acknowledge that he had even heard the prospect. When I asked him why he did this, he offered the following reasons:

(1) He doesn't want to lose momentum, or the flow of his presentation.

(2) Often, the objection will be addressed later as part of the presentation.

(3) Hearing the same objection twice means that it's important to the prospect and needs to be answered at that time.

(4) *Over half of unanswered objections disappear on their own. Thus, no response is necessary.*

One of my favorite responses is, "That's a good question. However, you're on page five and I'm on page two. I'll be explaining that in a few minutes."

Deal with a valid objection that you cannot directly overcome. This is one lesson I'll never forget. It cost me a huge commission to learn. I was running a region on the East Coast for a major educational marketing company. We had about forty salespeople and six sales managers in the region. I reserved a small territory to work myself just to stay fresh and in touch. I had set up a meeting with a fairly large school district in Eastern Pennsylvania. The entire decision-making body attended, including the superintendent and all the principals, about twelve people in total. The presentation was going very well, with plenty of positive feedback. However, I began to notice one principal in the back of the room sitting well apart from the group. He was a large man, stoic, almost unfriendly. While the "Yes" momentum from the other attendees continued to be reassuring, the loner in the back hadn't said a word, and I was beginning to feel a bit intimidated by his cold stare.

Suddenly, in a deep booming voice that filled the room, he unleashed a valid objection that I couldn't overcome. The man definitely exhibited the typical bully personality. The momentum began to shift. The other attendees became somber as he, leaning forward, restated his objection and expanded upon it as if we were competing. The others stopped supporting my presentation as the entire group turned negative. It was a classic "one-eighty." Needless to say, I did not get the sale.

I learned first hand, then and there, that one loud, negative voice in a group can easily become a majority. I stood there, feeling like an idiot. It was the first time in years that I was totally knocked off balance during a sales presentation. I had believed that there was nothing that could throw me like that in a selling situation. It was a very long ride home. I kept playing the incident over and over in my mind.

I called the VP of sales that evening and recounted the details of the meeting to him, expressing my utter frustration at having been totally decimated by one objection.

He said, "So, you don't know how to handle a loud, negative voice? I'll tell you how to do it. You agree with the person and you say, 'That's a good point. You're correct, we can't do that. (Pause) However, in our quest for perfection, let's not lose sight of all the benefits of our product.' Then, do a complete summary of all the benefits. As you build the list, the objection begins to shrink."

I wanted to get into my car and drive back to the school district. Obviously, the moment was lost. It was an expensive lesson, but it has paid great dividends over the years.

Sometimes an objection plays into your hands. Sometimes the prospect will pose an objection that can be easily handled by responding with, "That's the very reason you should be utilizing our service." Here's an example. The prospect says, "We really can't afford this." You reply, "That's the very reason you need to own this product. It will save you money on maintenance, energy usage, etc. In the end, it will pay for itself."

Uncover the hidden objection. Sales are lost when the prospect has an objection that hasn't been revealed. This is definitely not a rare occurrence. There are numerous reasons for this difficult situation. Perhaps there's a credit problem that would be embarrassing for the prospect to discuss. It could be that he/she is not the decision maker and doesn't want to admit it. Maybe the prospect really can't afford it and is ashamed to say so. Perhaps the prospect doesn't qualify for legal reasons. Maybe he/she doesn't trust or like the salesperson. There could be any number of additional reasons for prospect resistance.

The point here is that you cannot close a sale if there's a hidden reason preventing the prospect from making a decision. The challenge is to recognize that there is a hidden objection and to uncover it.

How do we know when there's a hidden objection? We can't be certain until it's exposed. However, we can get some reliable hints when the prospect is resisting and it doesn't make sense to you. This tactic is also effective when the prospect is not forthcoming with information that would reveal his/her readiness to buy.

The solution This is a rarely utilized, yet effective, tactic revealed to me by a legend in the encyclopedia sales business. He claimed that I would probably never read it in a book about selling or hear it in a sales training seminar. It's so simple that it went straight over my head the first time I heard it!

Ask the prospect to tell you why he/she's not going ahead with your proposal. "Please be candid with me and tell me why you're hesitating?"
(Shut up and listen).

(1) Once the prospect has given you a reason, you want to check it out to make certain that it's the real reason and not a fabrication. Therefore, remain silent for about five to seven seconds to

give the prospect the opportunity to expand upon the reason or add another one.

(2) Show empathy by repeating it in your own words as if you've accepted it. "In other words, you feel that _____ ." Then, remain silent for another five seconds.

(3) When you have waited long enough to feel that the prospect has nothing more to add, pose the key question: **And, what's your other reason?"**

The prospect might say that there is no other reason. At that point, you can be reasonably certain that the first reason is authentic. However, if the first reason is a fabrication, the hidden objection will fall onto the table like a pearl. This happens because you've probably extracted all the fabrication from the prospect and there is nothing left to offer but the truth. This tactic is simple to perform when delivered systematically. It requires timing and patience. You'll be amazed by its effectiveness.

Eight Rules for handling objections

(1)   Hear it as question.

(2)   Validate the prospect, "That's a good point."

(3)   Pause and reflect before responding.

(4)   Repeat it in your own words.

(5)   Never guess at an answer.

(6)   Never argue with the prospect.

(7)   Look for a closing opportunity each time.

(8)   Never accept an objection as a final "No."

Handling objections skillfully is just one of the tools that you need to be successful. Executing sound selling skills puts you in a position to avoid costly mistakes, control the variables, and increase your chances for a positive outcome.

LESSON 6

Closing the Sale

Sometimes, the prospect agrees to close on his own before the salesperson has a chance to ask. I call this phenomenon the **"pure sell."** The prospect just sort of takes it away from you: "How do we get started?" or "Let's do it."

This welcome event usually occurs when the presentation itself anticipates the objections and answers them as part of the presentation before they come up. The pure sell is the ultimate reward for a job well done – a perfect report card. It's about achieving all the objectives. It's the mark of true, professional salesmanship. The better you're prepared, the better you execute, the more often the pure sell occurs. It's a thrill each time it happens to me. However, it would be a fatal mistake to depend on the pure sell scenario to earn a living.

Buyers are becoming better informed and more value conscious at a faster pace than any time in history. They're interested in what happens before the close. They will focus on your research and how you meet their needs rather than your closing techniques. Does this reduce the need for sound closing skills? Absolutely not! It increases the need for more astute skills in all phases of the selling process, particularly the closing.

The major prerequisite to a successful close is that there is no misunderstanding on the prospect's part. When the buyer is unclear about the key points or any of the details that are important to him/her, you are likely to meet resistance. **Take whatever steps necessary to confirm that the prospect has not misunderstood.** This is one of the reasons that we summarize the key points after the presentation. This is why we ask the prospect several times throughout the presentation if he/she has any questions: "Are you with me?"

My practical approach to closing has always included the concepts taught by the sales giants of this field, such as J. Douglas Edwards, the modern father of sales instruction and closing, in particular. I was fortunate enough to attend one of his events. His presentation was responsible for elevating my career far more than anything I can remember. His body of work strongly influenced several of the closes found in this section.

We start here by defining a closing question. I have never been able to find a more accurate definition than the one presented by J. Douglas Edwards: "A closing question is any question you ask the prospect, the answer to which confirms the fact that he has bought." (Or not bought). Edwards then asked the audience for examples of closing questions.

He then offered the critical instruction: "Whenever you ask a closing question…

Shut Up!"

Why should you shut up? First of all, when you ask someone a question, common courtesy dictates that you provide as much time as necessary for the person to consider the question and to respond. It's rude to start talking before they've answered your question. Secondly, if you remain silent, there is only one of two things they can do – agree to your offer or let you know why they aren't buying. Many salespeople feel the pressure themselves and break the silence. Once the silence is broken by the salesperson, the moment is lost, and often the sale as well.

Eight Favorite Closes

Fill Out the Agreement Close

Prior to initiating this close, ask the prospect this, "Do you have any questions or concerns?" A "No" answer means "Yes, you may proceed."

This tactic is the single most effective way I know of to close a sale. You bring out the contract/agreement (which you have readily available, visible to the prospect and in your control).

You: "I have a few questions. Your address is?" (Write his answer on the form).

Fill in the answers. Keep asking for all the information until you've filled in every required space. If they stop you, you have more selling to do. If they don't, they're buying.

Turn the agreement around after you've filled in all the blanks, hand the pen to the prospect and say, "I need your approval here," while pointing to the signature line(s).

The words, "authorize this," "approve this," and "OK this," are softer words than "sign" and "signature." While they mean the same thing, some people might object to "signing" an agreement. They may not have a problem "approving" it.

The Two-Choices Close

This is a very commonly utilized, assumptive close. You give the prospect two choices. Once the prospect voices his/her decision, the deal is closed. Just write it up.

You: "Would you prefer to lease the equipment or make a cash investment?"

Prospect: "Lease."

You: "That's a good choice. I'll write it up."

Respond To a Single Question with a Close

This one is a straightforward, easy-to-articulate close, and it works most of the time. You simply answer a question that the prospect asks with a closing question. It goes like this.

Prospect: "Can you deliver it in thirty days?"

You: "Would you like to have it in thirty days?"

Prospect: "Yes."

You: "Great!" (It's closed. Fill out the agreement as described above).

Most salespeople would answer: "Yes, we can deliver in thirty days."

I've witnessed similar question and answer sessions that don't seem to have an ending, Numerous "Yes" responses are given, but zero closing attempts are made.

Special-Treatment-Request Close

I suggest that you use this close when the prospect is asking for special treatment that requires extra effort on your part but you are reasonably certain you can meet the request or get close to it, e.g., terms, pricing, irregular changes, additions, or deletions.

Prospect: "Can you give us the right of first refusal with the agreement?"

You: "Quite frankly, I'm not authorized to do something like that. I'll have to run it by our legal department. Let me ask you this before I check it out. If we can provide you with the right of first refusal, can we finalize this agreement?"

Prospect: "Yes." (If the answer is "No", you have more work to do.)

You: "Fine, I'll add it to the agreement right here; an addendum stating that this agreement is 'Subject to providing you with the right of first refusal,' I need you to initial it here. We have a deal if it's accepted; fair enough? "

Put-It-All-In-Writing Close

This close is effective when the prospect is on the fence, hesitating and needing reassurance.

You: "Can I make a suggestion, please?"

Prospect: "OK"

You: "I'm going to put everything down in writing here, so you can see that we haven't left anything out. We can go over it together to make certain that you're getting everything promised. I want you to be comfortable with this. Then, you can approve it and move forward with peace of mind; fair enough?"

Hold your hand out to invite a handshake and seal the deal. The handshake is important. It provides the tactile assurance that adds the final touch.

Give-Me-a-Rating Close

There are times when we've done everything right and still don't have a clue where the prospect stands. This one works well with the stoic, non-responsive prospect.

It has proven to be effective at any level, including long-term negotiations for huge stakes.

The objective is to take the prospect's temperature and close the deal in one fell swoop. I've closed some very difficult sales with this one. It's simple, straightforward, and mighty powerful.

You: "Mr./Ms. Prospect, time out for a moment.
(Pause) How are we doing on this? I need you
to let me know where I stand. Give me a rating
from one to ten, ten being the best."
(Pause, wait for a response).

Prospect: "Seven."

You: "Seven? I didn't think I was doing that well.
What's it going to take to get to ten?"

It doesn't matter what number the prospect gives you.
Use the same response, unless it's a "10," in which
case, close it.

The prospect should then tell you everything that
needs to be done to close the deal. You have the
option to go directly into the "hidden objection"
scenario if you feel that the prospect is not being fully
candid with you.

Ask for the Business

This is a direct, easy-to-deliver request. You've gotten
positive feedback. Now, simply ask: "Can we go forward
with this?" (Shut up.)

The Humorous Close

Sometimes a bit of humor can be an effective tool during the closing process. I was managing a salesman who was becoming very frustrated with a prospect who simply wouldn't respond. He was a Japanese gentleman, president of a large insurance agency, very stoic, the inscrutable Asian stereotype.

After reviewing our proposal, it was easy for me to see that the proposition made a great deal of sense for his business. His company was utilizing an antiquated telephone system. Leasing our advanced equipment would significantly improve communications and save his agency a considerable amount of money. I decided to join our salesman for his fourth visit.

His receptionist respectfully ushered us into his huge, impressive office. We sat at his desk as he stared at me. I knew that I had to do something different, perhaps risky, to get control of the situation.

I immediately got to the point. "I'm here today to make you an unusual offer as an inducement to get you to act on our proposal. I've never offered this to anyone else as long as we've been in business." Then, I shut up and waited. I was determined not to break the silence. Finally, curiosity got the best of him and he asked, "What is it?" I waited a few beats before replying, "If you're willing to go along with us today with our phone system, I'm prepared to completely forgive you for Pearl Harbor."

The expression on his face didn't change for several moments. Suddenly, he broke out laughing and so did we out of a sense of relief. Then, with a wide smile on his face, he said, "I never do business with anyone unless they have lunch with me." Needless to say, we closed the deal.

LESSON 7

Trial Closing

Trial closing is a softer, less confrontational technique than straight, direct closing questions. The objective is to test the prospect's readiness to buy, in that it asks the prospect for an opinion. A closing question asks for a decision, in which case the prospect can reply with a flat out "No." At that point, the odds of success have diminished and one has much more selling to do to recapture positive momentum. I seldom move directly to the close until I've asked several trial closing questions throughout the presentation. There are numerous opportunities during the sales process to trial close.

Examples of Trial Closing Questions:

"Where would you house the main controller so it would be safe and out of sight?"

"Tell me how you'd decorate the living room."

"When do you think would be the best time to announce the acquisition?"

"When is your target date for moving?"

"Would a two-week notice for training be sufficient?"

"Most of our clients really like the no-interest option. How would that work for you?"

You will rarely hear the word, "No" when asking trial closing questions.

Trial closing will give you the clues you need in order to know when to close and will contribute to higher percentage selling.

LESSON 8

Fear of Failure and Rejection

The fear of failure and rejection are closely related and are the biggest impediments to successful selling. The word "No" is one of the first negative words we hear as very young children. It's often delivered with a harsh, abrupt, urgent, and stinging voice. It's a word that often robs us of what we're doing, where we're going, and what we want. Sometimes we feel that we have failed to please when we hear "No". Perhaps we are scolded and made to feel ashamed by our parents or peers when we hear the word. It doesn't take long to learn that it's much safer to avoid the word "No", as we associate it with negative events and feelings. Thus, we avoid it. We become afraid of it. And the smarter one is, the more ways one can devise to avoid it.

What is the fear of rejection? Why does it negatively impact our ability to become good at selling?

Fear of rejection is the

- **Irrational fear** that others will not accept me for who I am, what I believe, and how I act. It is the pervasive motivator for caution in my behavior and interactions with others.

- **State of mind** that makes me incapable of doing or saying anything for fear of others' rejection, lack of acceptance, or disapproval.

- **State of being of individuals who are over-dependent on the approval,** recognition, or affirmation of others in order to feel good about themselves. In order to sustain personal feelings of adequacy, these individuals are constantly concerned with the reactions of others to them.

- **Self-censoring attitude** that inhibits creativity, productivity, and imagination.

- **Driving force** behind many people that keeps them from being authentic human beings. They are so driven by the need for acceptance of others that they lose their own identity in the process. They mimic the ways in which others act, dress talk, think, believe, and function. They become three-dimensional clones of the role-models they so desperately need to emulate in order to gain acceptance.

- **Underlying process in the power of "peer pressure"** that grabs hold and makes people behave as a stereotype, such as pop culture, counter culture, punk, new wave, preppie, yuppie and other styles. They crave recognition and acceptance from the reference group with whom they want to be identified.

- **Energy-robbing attitude** that leads to self-immobilization, self-defeating, and self-destructive behavior. This attitude encourages ongoing, irrational thinking and behavior resulting in personal stagnation, regression, and depression.

- **Driving force for some people for all actions in their lives.** It plays a part in their choices concerning education, career direction, work behavior, achievement level, interpersonal and marriage relationships, family and community life,

and the ways
in which they spend leisure time.

**Act of giving to others more power than I give
myself** over how I feel about myself. What the others
say or feel about me is the determinant of how I feel
about myself. I am completely at the mercy of others
for how happy or sad I will be. My self-satisfaction and
belief in myself is in their hands. Fear of rejection is
the abdication of power and control over my own life.

If you can say, "Yes, this is me," to more than two of
the above items, you probably have some inside work
to do.[1]

Everyone feels some degree of this fear. I know that
it interferes with quality of life and becomes a barrier
to healthy communication. I also know that the sales
greats have come to terms with it and have mastered
it. Perhaps, it's their preparation that reduces the fear.

I have no problem acknowledging that I have a
fear of rejection. I don't hide from it. I do my best to
walk through it whenever it threatens me, always to

---

1   The information, "What is fear of rejection", was quoted from **www.coping.
org** "fear of rejection." Visit the site above for more helpful information.

discover that the results aren't as bad as I feared. The closer you get to your fears, the smaller they become.

At the end of the day, the ultimate question is, "Which is going to prevail, your need to succeed or your fear of failure?"

Each of us lives in our own cocoon of habits and beliefs. It's comfortable in there. The thought of giving up the cocoon is scary. We don't quite know what life would be like out there. However, you must step outside of your comfort zone and take some perceived risks if you really want to grow.

"Fear hides behind indecision"

Unknown

Visualization – The Mental Advantage

If there is only one thing that you adopt and utilize in this book, let it be this lesson: learn the simple exercise of visualization. It is a proven method for achieving dreams and materializing goals.

> *"The subconscious mind is the sum total of our past experiences. What we feel, think, or do forms the basis of our experience. These experiences are stored in the form of subtle impressions in our subconscious mind. These impressions interact with one another and give birth to tendencies."*
>
> C.S. Shah

In simple terms, the subconscious mind is the repository of the mind that has stored every single thing in your life that you've seen, heard, done, and experienced. It is the floor plan for your identity. It is the source of your dreams while sleeping. It's where your thoughts come from when driving down the highway on cruise control without anything pressing on your mind. It feeds the conscious mind.

Here's the secret about the subconscious mind. It has no prejudice. It accepts everything. *It cannot distinguish between an imagined event and a real happening.* This quality, in itself, is the basis of the power visualization offers. Sports psychologists have done numerous experiments with visualization and have recorded some amazing results. Many sports teams, both at the college and the professional levels, utilize it effectively.

"I took the visualization process from the Apollo program, and instituted it during the 1980's and '90's into the Olympic program. It was called 'Visual Motor Rehearsal.'

"When you visualize then you materialize. Here's an interesting thing about the mind: we took Olympic athletes and had them run their event only in their mind, and then hooked them up to sophisticated bio feedback equipment. Incredibly, the same muscles fired in the same sequence when they were running the race in their mind as when they were running it on the track. How could this be? Because the mind can't distinguish whether you are really doing it or whether it's just a practice. If you've been there in the mind, you'll go there in the body."

Dr. Denis Waitely,
From the book, *The Secret* by Rhonda Byrne.

Visualization works! I know because I've been using its power for many years.

It's easy. You can do it. And, it works.

Here are the steps to make it happen:

(1) Write a brief, one-paragraph scene of you being successful at whatever you wish to achieve. It can be almost anything. In the case of successful selling, it could be a scene of you receiving a sales award while being honored by your peers.

It might it be a scene of you writing several huge orders or giving a great presentation, etc.

(2) Once a day, preferably before you fall asleep, close your eyes and imagine the scene in your mind's eye. Simply view the short scene you've written as if you are watching a film.

(3) Repeat this exercise daily at about the same time. If you miss, it's OK. Do it first thing in the morning. Make this viewing a "must do" every day!

Here's what's going to happen: Your subconscious mind will begin to deliver activity directions to your conscious mind to achieve what you've been viewing in your film. You'll probably find yourself doing things that you haven't done before, or haven't done with such consistency and determination. Keep watching yourself in the act of being successful. You will become successful!

Keep Earl Nightingale's words ever present in your mind: *"You become what you think about."*

A Personal Story

It was over twenty years ago that I went "flat out" broke. I had owned fifty percent of a telecommunications company as the vice president of sales. I moved sales from less than $400,000 the first year to over $10,000,000 in year six. I actually thought that I was ten feet tall, invincible, and could fly. We had four full-blown operations, each in a major city, with over eighty employees. We had grown the company beyond our ability to manage it. Our credit line was fully tapped out. The company began to lose money; my partner and I ceased taking salaries. Shortly thereafter, the company failed. I had a new wife and a new baby. We were living in an affluent community in Northern California. We had live-in help. We had to let them go.

I suddenly realized that we were broke, zero income. For the first time in my life I knew what the threat of losing everything meant. We didn't have money to pay our mortgage or buy groceries. My wife, Sharon, with tears in her eyes, asked, "Am I going to be a bag lady?" It was the darkest day in my life.

The creditors and the sharks took every last shred of what was left of the company, with the exception of one demonstrator business telephone. It was a sobering experience, to say the least. Luckily, we escaped without having to declare corporate or personal bankruptcy.

I took inventory and recognized that I was left with four assets of value: the title to Sharon's car, which I promptly pledged to borrow $5,000 from the bank, one demonstrator telephone, great selling skills, and experience. I knew that I had to crawl out of a place I had never been, a dark, deep hole. I knew that I would succeed if I could keep my bearings and do the things that I knew how to do. My options were, indeed, very limited. I could not afford the luxury of feeling sorry for myself or indulging in negative thinking.

I spent the next two days going through the Yellow Pages and phoning companies to identify those that would be viable prospects to purchase a new business telephone system. I asked for the president's name each time. Then, I began the process of dialing for dollars. By the end of the week, I had made four appointments for the initial meeting with four presidents of companies, out of the list of about forty prospects.

I utilized an industry selling standard, a five-step selling approach. After the end of the second week, I had completed four of the steps in each company and had made appointments with each of their presidents for the final step, the proposal presentation and close. I must admit, I was very nervous.

The first presentation was to the local president of a large, international bank with a huge branch in the financial district of San Francisco. He sat motionless as I presented the features, benefits, and value proposition of owning the phone system. Upon completion, I simply shut up and waited for him to speak. Finally, he asked, "How long have you been in business?"

I waited a couple of beats, and looked at my watch without a verbal response.

He paused for a moment and replied, "You are kidding aren't you, How many customers do you have?"

I answered, "You're going to be the first one."

That one stopped him in his tracks momentarily as he processed the information. I kept looking at him with a slight smile on my face. I sat perfectly still.

Then he asked a question that I wasn't quite prepared to answer, "Why should I do business with you? You're asking me to place the nerve center of our bank in your hands. This is an international bank! You're just starting in business and you have no customers. Why should I do business with you?"

I leaned forward slowly, as if to tell him a secret and lowered my voice, "You probably shouldn't. I don't know how you could explain it to your associates, your attorney, or anyone; (pause) but, if you do go along with me, you will own me for life."

I went on to tell him about my experience in the telecom business and brought him up to the present moment. He sat silently for what seemed like an eternity. So did I. He picked up his phone and asked his assistant to bring in a check for fifteen thousand dollars, which represented the required thirty percent deposit for a fifty-thousand-dollar telephone system. Then he cupped his hand over the mouthpiece and asked, "What's the name of your company?" I told him, and he told his assistant to whom the check should be made out.

Several minutes later, his assistant arrived and handed him the check.

He signed it and slid it across the table in front of me. I didn't look at it. I let it sit there for the longest time as I struggled with myself not to pick it up too soon. We discussed installation procedures, training dates, etc. He then suggested a day at the races and I accepted.

Once I was alone in the elevator, I broke down. The next three presentations that week ended with a contract and a check. I had funded my new company with customer deposits. I soon rented space, hired some people, and was off and running again, this time with a little more experience. I sold that company four years later for a considerable amount of money. Sharon never became a bag lady.

About the Author

 Gerry Shaltz introduces his first book, *The DNA of Selling,* to share the wealth of knowledge he has gained over many years. His experience in sales, managing, training, and lecturing has earned him much deserved respect in the business community. Gerry is an entrepreneur and multi-faceted businessman. He has a refreshing awareness and uncommon wisdom that he has shared in lectures and workshops with graduating MBAs at major universities, companies, and business organizations.

In addition to his keen understanding of selling and managing salespeople, he has been highly successful working with start-up companies. He is the co-founder and an officer of Seismic Warning Systems, Inc., a company that has developed the world's leading technology for early warning earthquake systems. He also served on the board of a Beta Group company, Beta Frames, LLC.

The author's success evidences his ability to use his proven methods, instincts, skills, and gifts to build lasting relationships with clients while doing what he truly loves. Mr. Shaltz has never wavered in his passion for selling and for teaching others how to succeed at selling.

To contact Gerry Shaltz:
www.gerry@thednaofselling.com

To purchase additional copies:
www.amazon.com